FLORA OF TROPICAL EAST AFRICA

HALORAGACEAE*

R. Boutique
(Jardin botanique national de Belgique)

and

B. Verdcourt
(Royal Botanic Gardens, Kew)

Aquatic or terrestrial herbs or subshrubs with hermaphrodite, monoecious or dioecious flowers. Leaves usually exstipulate, alternate, opposite, whorled or all radical, sometimes very small, the blades simple, entire or ± divided especially in the case of submerged leaves. Inflorescences axillary or terminal, the flowers in cymes, fascicles, racemes, panicles, spikes or solitary; bracts often present. Flowers often bracteolate, usually small, regular. Calyx with the tube adnate to the ovary; lobes 2–4, valvate, sometimes rudimentary or absent. Petals 2(–3)–4, valvate or slightly imbricate-contorted, or sometimes absent. Stems (1–)2–8; filaments long or short; anthers basifixed, 2-thecous, dehiscing by lateral slits. Ovary inferior, rounded or angular to winged, 1–4-locular; locules 1-ovulate; ovule apical, pendulous, anatropous; styles 1–4, free or absent; stigmas papillate or plumose. Fruit a nutlet or drupe, bluntly angular, ribbed or winged with 1–4 1-seeded locules or divided in 2 or 4 1-seeded cocci. Seeds with abundant fleshy albumen.

A small family of 7 genera and about 130 species occurring in temperate, subtropical and tropical regions; in the Flora area there are 2 native genera with but 1 species each and 1 introduced genus with 2 species. One of these introduced species appears to have escaped sufficiently widely for it to be treated fully. An account of the palynology of the family is given by Praglowski in Grana 10: 159–239 (1970).

Plants not floating in water; leaves not finely divided:
  Plants acaulescent; leaves very large and long-petiolate,
    borne on the rhizome; inflorescences borne on the
    rhizome . . . . 1. **Gunnera**
  Plants with stems; leaves generally very small, sessile
    or shortly petiolate; inflorescences axillary. . 2. **Laurembergia**
Plants floating in the water; leaves in whorls, the sub-
  merged ones at least divided into filiform segments 3. **Myriophyllum**

1. GUNNERA

L., Mant. Pl.: 121 (1767) & in Syst. Nat., ed. 12: 597 (1767); Schindler in E.P. IV. 225: 104 (1905); Bader in Webbia 19: 537, fig. 1 (1965) [world distribution]; van der Meijden & Caspers in Fl. Males. 7: 259 (1971)

Perennial stemless herbs with a creeping or erect rhizome and sometimes with dioecious flowers. Leaves borne on the rhizome, often very large, some-

* Sometimes and perhaps more correctly spelt Haloragidaceae but conserved with the spelling used here. This account has been translated from that of R. Boutique for the Flore du Congo, du Rwanda et du Burundi by B. Verdcourt with some necessary additions and alterations made.

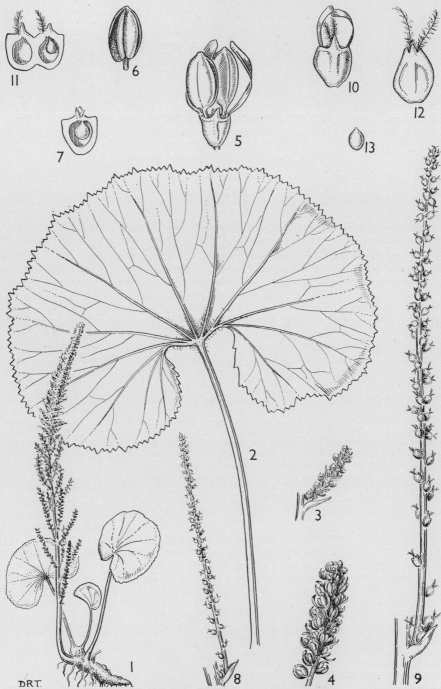

FIG. 1. *GUNNERA PERPENSA*—1, habit, × ⅓; 2, leaf, × ½; 3, spike of male flowers, × 1; 4, same, × 2;
5, male flower, × 8; 6, stamen, × 8; 7, section of ovary from male flower, × 8; 8, spike of female flowers,
× 1; 9, same, × 2; 10, female flower, × 8; 11, ovary from same, opened out, × 8; 12, older female
flower, × 1; 13, ovule, × 8. All from *St. Clair-Thompson* 799. Drawn by Miss D. R. Thompson.

times provided with stipule-like appendages; petiole equalling the limb or much longer; limb suborbicular, reniform or ovate, ± deeply cordate at the base, lobed, crenellate or dentate with glandular teeth, rarely entire. Inflorescences borne on the rhizome, spicate, racemose or paniculate, composed either of unisexual flowers or of distinctly protandrous ☿ flowers or of ☿, ♀ and ♂ flowers, the latter being apical; bracts present or absent. Hermaphrodite flowers with calyx-tube ovoid or ± compressed, with (0–)2(–3) thick lobes; petals 2, small, or absent; stamens 2; ovary 1-locular; styles 2, filiform, subulate or compressed, papillose. Male flowers with calyx-tube reduced; styles none or rudimentary. Female flowers without petals. Fruit a nutlet or drupe, almost globose or 3-angled, coriaceous to somewhat fleshy. Seed 1, with a membranous testa.

A moderate-sized genus of 40–50 species in S. and tropical Africa, Madagascar, Indonesia, Tasmania, New Zealand, Antarctica, Hawaiian Is., Central and South America. The single species occurring in the Flora area belongs to subgen. *Gunnera* (subgen. *Perpensum* (Burm.) Schindler).

G. perpensa *L.*, Mant. Pl.: 121 (1767) & in Syst. Nat., ed. 12: 597 (1767); Sims, Bot. Mag. 50, t. 2376 (1823); Harv. in Fl. Cap. 2: 571 (1862); P.O.A. C: 297 (1895); Schindler in E.P. IV. 225: 116 (1905); T.C.E. Fries in N.B.G.B. 9: 36 (1924); Staner in Ann. Soc. Sci. Brux., sér. 2, 59: 138 (1939); Boutique, Fl. Congo, Haloragaceae: 2, t. 1 (1968). Type: South Africa, Cape of Good Hope, *Burman* in *Linnaean Herbarium* 1063.1 (LINN, lecto.!)

Herb 15–100 cm. tall. Rhizome often deep red or yellow, up to about 2·2 cm. thick. Leaves with robust petiole 15–80 cm. long, sparsely pubescent to glabrous; limb reniform to almost round, broader than long, 4–15(–24) cm. long, 5·5–26·5(–30) cm. wide, cordate at the base, toothed, pubescent to quite densely almost scabrid-velvety hairy with stiff adpressed hairs particularly on the nerves and in the young state, becoming glabrescent. Spikes solitary or fasciculate, sparsely to densely arranged along an axis (15–)25–100(–150) cm. long of which the peduncle, often red at the base, comprises 2–35 cm.; spikes 0·2–13·5 cm. long, the basal ones the longest, many-flowered, sparsely pubescent to glabrous; bracts elliptic-lanceolate to linear, 4–8 mm. long, ± pubescent or glabrous. Flowers ♂, ♀ or ☿, pink or pale violet; calyx-tube green, 0·5–1·5 mm. long, lobes reddish, triangular, minute; petals reddish brown, ± linear-spathulate, 2–3·6 mm. long, glabrous or hairy at the apex, sometimes absent; anthers 1–1·9 mm. long; styles subulate, plumose, (0·4–)0·9–1·4 mm. long, or rudimentary or absent. Fruits sessile, subsessile or slightly pedicellate, globose, 1 mm. in diameter, glabrous, with bases of calyx-lobes and stamens persistent. Fig. 1.

UGANDA. Kigezi District: Bufundi, July 1946, *Purseglove* 2101! & same locality, 23 Dec. 1933, *A. S. Thomas* 1204!
KENYA. W. Suk District: upper reaches of R. Morun, Mar. 1965, *Tweedie* 3014!; S. Mt. Kenya, R. Ruamuthambi [Ramusambi] source, Kiamariga [Kilimarigo] Swamp, *Gardner* in *F.D.* 1882! & Mt. Kenya, Marania R., 14 Feb. 1922, *Fries* 1543!
TANGANYIKA. Masai District: Ngorongoro Highlands, between Munge and Oldonyo Was, 22 Sept. 1932, *B. D. Burtt* 4296!; Njombe District: Kipengere Mts., Nyarere R., 9 Jan. 1957, *Richards* 7624!; Songea District: about 1·5 km. N. of Miyau, by R. Utili, 2 Mar. 1956, *Milne-Redhead & Taylor* 8950!
DISTR. U2; K2–4; T2, 3, 7, 8; Zaire, Rwanda, Burundi, Sudan, Ethiopia, Rhodesia, Mozambique, South Africa, Madagascar
HAB. Swamps, damp ground, thick grass by streamsides, etc.; 1560–3800(–4000) m.

SYN. *G. perpensa* L. var. *kilimandscharica* Schindler in E.P. IV. 225: 117 (1905); Z.A.E.: 589 (1913); T.C.E. Fries in N.B.G.B. 9: 36 (1924); Boutique, Fl. Congo, Haloragaceae: 4 (1968). Types: Tanganyika, Kilimanjaro, *Engler* 1830, *Volkens* 749, 925 & 2097 & W. Usambara Mts., Mbalu, *Engler* 1445 & *Buchwald* 346 (B, syn. †)

G. perpensa L. var. *angusta* Schindler in E.P. IV. 225 : 117 (1905). Type:
Madagascar, S. Betsileo, Ankafira Forest, *Hildebrandt* 3962 (B, syn, †, K,
isosyn. !) & without locality, *Baron* 2238 (G, syn., K, isosyn. !)
G. perpensa L. var. *alpina* T.C.E. Fries in N.B.G.B. 9 : 36 (1924). Type: Kenya,
Aberdare Mts., alpine region, 3800 m., *Fries* 2464 (UPS, holo., K, iso. !)

NOTE. I do not believe that the varieties mentioned above are worth keeping up.
Var. *kilimandscharica* is supposed to differ from the typical variety in its glabrous
petals and pedicellate fruits. I have examined a good deal of material from Kilimanjaro and none has the pedicels really developed; the petals vary greatly in hairiness
and there is only a tendency for the S. African material to have more hairy petals.
Schindler recorded var. *kilimandscharica* from tropical Africa and var. *perpensa* from
South Africa but Boutique records both from tropical Africa, the var. *kilimandscharica*
from Kilimanjaro and Zaire (Kivu Province), although he had not seen the material.
T. C. E. Fries refers *Volkens* 925, a syntype of var. *kilimandscharica*, to his var. *alpina*
in an informal way and Schindler mentions that *Volkens* 2097 and 925 represent a small
alpine form.

2. LAUREMBERGIA

Bergius, Descr. Pl. Cap.: 350 (1767); Schindler in E.P. IV. 225 : 61 (1905);
A. Raynal in Webbia 19 : 683–695 (1965); van der Meijden & Caspers in Fl.
Males. 7 : 246 (1971)

Herbs, sometimes somewhat suffruticose, with a woody, ± creeping
rhizome and rooting at the lower nodes; stems often reddish. Leaves opposite,
verticillate or alternate, generally small, sessile or shortly petiolate, the limb
entire or lobed. Inflorescences axillary 1–11-flowered fascicles, sometimes of
1–3 pedicellate ♂ flowers and the others ♀ and sessile or subsessile, or sometimes 1 long-pedicellate ♂ flower and the others ♀ and sessile or subsessile, or
sometimes long-pedicellate ♂ flowers in the axils of the upper leaves and ♀
sessile or subsessile flowers in the axils of the lower leaves. Flowers with
calyx-tube ellipsoid or urceolate, with longitudinal nerves and also longitudinal often strongly mamillate ribs; calyx-lobes 4, persistent; petals 4,
sometimes rudimentary or lacking in female flowers; stamens 4 or 8; ovary
4-locular, becoming 1-locular through the dissolution of the septa; ovules 4;
styles 4 or absent; stigmas plumose. Nutlets very small, ribbed or not. Seed
1, pendulous.

A small genus considered by van der Meijden & Caspers to contain only 4 species
(Raynal estimated 10) in tropical and subtropical Africa, tropical Asia and America.
The sole species occurring in the Flora area belongs to the subgenus *Serpiculastrum*
A. Raynal, and is very variable.

L. tetrandra (*Schott*) *Kanitz* in Fl. Brasil. 13 (2): 378 (1882); Schindler in
E.P. IV. 225 : 74 (1905); A Raynal in Webbia 19 : 693 (1965); Mendes in
C.F.A. 4 : 31 (1970). Type: Brazil, *Schott* (?BP, holo.)

Suffrutescent semi-decumbent or ± erect herb 10–20 cm. tall, sometimes
forming mats; stems often reddish, glabrescent to entirely and densely
pubescent-villous. Leaves opposite or alternate, sessile or shortly petiolate,
slightly fleshy; blades linear, linear-oblong or narrowly ovate, elliptic or
obovate, 0·5–1·5 cm. long, 0·5–7 mm. wide, rounded at the apex, cuneate at
the base, entire or slightly to distinctly 1–4-toothed towards the apex.
Fascicles of 7–15 flowers, of which 1 is pedicellate and ♂ and the others sessile
or subsessile and ♀. Hermaphrodite flowers with pedicel 0·3–1·8 mm. long;
calyx green or pinkish, with the tube slightly constricted at the apex, ±
0·3 mm. long, with 8 smooth or 3–4-mamillate ribs and ± distinct longitudinal
nerves between the ribs; calyx-lobes narrowly triangular, 0·2 mm. long,
acuminate; petals 4, white or pink, oblong, 0·8–1·2 mm. long, cucullate;
stamens 4; anthers linear-oblong; filaments filiform; styles present; stigmas

FIG. 2. *LAUREMBERGIA TETRANDRA* subsp. *BRACHYPODA* var. *BRACHYPODA*—**1**, habit, × 2 ; **2**, detail of flowering branch, × 8 ; **3**, hermaphrodite flower, × 20 ; **4**, female flower, × 60 ; **5**, fruit from hermaphrodite flower, × 30 ; **6**, fruit from female flower, × 30. Var. *MILDBRAEDII*—**7**, leaf, × 6. 1–6, from *Purseglove* 1745 ; 7, from *Michelmore* 909. Drawn by Mrs M. E. Church.

capitate. Female flowers without petals. Nutlets pink, obovoid, 0·7 mm. long, 0·5 mm. wide, with 8 smooth or 3–4 mamillate ribs often confluent in pairs, the tubercles white and bead-like.

SYN.   *Haloragis tetrandra* Schott in Sprengel, Syst. Veg. 4, app.: 405 (1827)

subsp. **brachypoda** (*Hiern*) *A. Raynal* in Webbia 19: 694, adnot. (1965); Boutique, Fl. Congo, Haloragaceae: 5 (1968); Mendes in C.F.A. 4: 31 (1970).   Type: Angola, Huila, Humpata, R. Quipumpunhime, *Welwitsch* 1621a (BM, lecto. !, COI?, LISU?, P, isolecto.)

Fruit with 8 ribs ornamented_with (2–)3–4 distinct tubercles.

DISTR.   U2, 4; K3; T1, 3, 4, 6–8; northern and tropical Africa, Madagascar and Mauritius

var. **brachypoda**; A. Raynal in Webbia 19: 694 (1965); Boutique, Fl. Congo, Haloragaceae: 5, fig. 1/A (1968); Mendes in C.F.A. 4: 32 (1970)

Plants glabrous or glabrescent to densely villous-pubescent; leaf-blades linear-oblong to linear (ratio of length to breadth over 2·5), not discolourous.   Inflorescence 8–11-flowered; ♂ flowers with pedicel 0·5–0·7 mm. long; petals 0·8 mm. long. Fig. 2/1–6.

UGANDA.   Masaka District: Sese Is., Bukasa I., 27 Feb. 1945, *Greenway & Thomas* 7197 ! & Lake Nabugabo, Aug. 1935, *Chandler* 1320 !; Mengo District: Entebbe, 8 Feb. 1958, *Lind* 2321 !
KENYA.   Trans-Nzoia District: Kitale, 13 May 1953, *Bogdan* 3732! & same place, 14 Sept. 1958, *Napper* 790 ! & same area, prison dam, Mar. 1967, *Tweedie* 3423 !
TANGANYIKA.   Bukoba, June 1931, *Haarer* 2042 !; Mbeya Mt., 13 Dec. 1962, *Richards* 17043 !; Iringa District: Mufindi, Lake Luisenga, 15 Mar. 1962, *Polhill & Paulo* 1761 !
DISTR.   U4; K3; T1, 4, 7, 8; Senegal, Portuguese Guinea, Liberia, Ghana, Nigeria, Central African Republic, Zaire, Sudan, Malawi, Mozambique, Zambia, Rhodesia and Angola
HAB.   Grassy edges of swamps, sandy lake margins, on mud by swamps in grassland areas, pathsides, etc.; 780–2100 m.

SYN.   [*Serpicula repens* sensu Oliv., F.T.A. 2: 405 (1871), *non* L.]
S. *repens* L. var. *brachypoda* Hiern, Cat. Afr. Pl. Welw. 1: 332 (1896)
Laurembergia repens Bergius var. *brachypoda* (Hiern) Hiern, Cat. Afr. Pl. Welw. 2: 482 (1901)
L. *angolensis* Schindler in E.P. IV. 225: 72 (1905).   Types: Angola, Lubango, Huila, *Dekindt* 230 (B, syn. †, LUA, isosyn.) & R. Lopolo, *Welwitsch* 1621 (B, syn. †, BM, K, isosyn.!) & *Welwitsch* 1621B (B, syn. †, BM, COI, K, LISU, isosyn.!)
L. *engleri* Schindler in E.P. IV. 225: 73, fig. 21 (1905); F.W.T.A., ed. 2, 1: 171 (1954).   Types: Nigeria, R. Niger, Nupé, *Barter* 1665 (B, syn. †, K, isosyn.!) & Sudan, Bahr el Ghazal, Jur [Djur], Seriba Ghattas, *Schweinfurth* 2582 & 2582a (both B, syn. †, K, isosyn.!) & Angola, Quimbundo [Kibundo], *Pogge* 145 (B, syn. †)
L. *villosa* Schindler in E.P. IV. 225: 74 (1905); F.W.T.A., ed. 2, 1: 171 (1954). Types: Senegal, Mboro, *Perrottet* 742 & *Leprieur* (G, syn.)
[L. *tetrandra* (Schott)Kanitz var. *numidica* sensu A. Raynal in Webbia 19: 694 (1965), pro parte; Boutique in Fl. Congo, Haloragaceae: 6, fig. 1/B (1968), pro parte; Mendes in C.F.A. 4: 32 (1970), pro parte, *non* (Batt. & Trab.) A. Raynal sensu stricto]

NOTE.   Some specimens have been referred to var. *numidica* (Batt. & Trab.) A. Raynal (based on *Serpicula numidica* Batt. & Trab., Fl. Anal. et Synopt. Algér. et Tunisie: 128 (1905); type: Algeria, La Calle, *Durieu* Flora selecta exsiccata 836 (P, lecto.)), but to my mind the ± glabrous tropical African material, e.g. that from Uganda and Tanganyika cited above, is far closer to hairy var. *brachypoda* than it is to the glabrous var. *numidica* from N. Africa, Madagascar and the Mascarenes and I have hesitated to refer it to var. *numidica*. There are numerous intermediates between the glabrous and hairy forms of var. *brachypoda*; they have been seen from Sierra Leone, Ghana, Zaire, Sudan, Malawi, Zambia and Rhodesia apart from East Africa.   The name *L. engleri* Schindler has frequently been applied to the glabrous form and to these intermediates and indeed the syntypes of *L. engleri* are intermediates.   An example of an intermediate less hairy specimen from the Flora area is *Stolz* 2465 ! (Tanganyika, Njombe District, Ukinga, Madehani, 26 Jan. 1914).   The leaf-blades are nearly always linear but a Senegal specimen determined by Raynal has elliptic blades 1·6 × 0·8 cm.

var. **mildbraedii** (*Schindler*) *A. Raynal* in Webbia 19: 694 (1965); Boutique, Fl. Congo, Haloragaceae: 7, fig. 1/C (1968).   Type: Rwanda, Rugege, *Mildbraed* 965 (B, holo. †)

Plant generally glabrous or glabrescent but sometimes with the stems very densely pubescent; leaves sessile or with petiole up to 1·5 mm. long; blade ovate, subelliptic, or obovate, 0·5–1·3 cm. long, 2–7 mm. wide, ratio of length to breadth 1·5–2·5, rounded at the apex, cuneate at the base, entire or 1–4-toothed, often discolourous, sometimes ciliate and with a few hairs on the blade; inflorescences 7–15-flowered; ♂ flowers with pedicel 0·3–1·8 mm. long, sometimes pubescent; petals 1–1·2 mm. long. Fig. 2/7, p. 5.

UGANDA. Toro District: E. Ruwenzori, Mpanga R. source, July 1940, *Eggeling* 4022!; Kigezi District: 0·4 km. S. of Ruchetera P.W.D., Kashambya Swamp, 5 Sept. 1952, *Norman* 151! & Kabale, Butale, Nov. 1934, *Synge* S1211!
TANGANYIKA. Bukoba, Aug. 1931, *Haarer* 2142!; Pare District: S. Pare Mts., Tona, 11–12 Feb. 1915, *Peter* K 657! & K 658!; Morogoro District: Uluguru Mts., Lukwangule Plateau, 30 Jan. 1935, *Bruce* 716!
DISTR. **U**2; **T**1, 3, 6; Zaire, Rwanda and Burundi
HAB. Swamps (particularly *Sphagnum*), damp valleys or grassy plateaux; 1200–2550 m.

SYN. *L. mildbraedii* Schindler in F.R. 9: 124 (1911); F.P.N.A. 1: 685 (1948)
    L. androgyna Peter, *nom. nud.*; Raynal in Webbia 19: 694 (1965), *pro syn.*

DISTR. (of species as a whole). Eastern S. America, N. Africa (Algeria), tropical Africa, Mauritius and Madagascar

3. MYRIOPHYLLUM

L., Sp. Pl.: 992 (1753) & Gen. Pl., ed. 5: 429 (1754); Schindler in E.P. IV. 225: 77 (1905); van der Meijden & Caspers in Fl. Males. 7: 248 (1971)

Perennial, rarely annual, aquatic herbs, either free-floating or with rhizomes rooted in the bottom; foliage submerged, apart from the inflorescences (except in unusual terrestrial forms). Leaves in whorls of 3–6, or aerial ones whorled, opposite or alternate, without stipules or with 1(–3) filiform to subulate deciduous stipule-like outgrowths; submerged ones pinnately divided into unbranched capillary segments; aerial ones sometimes simple, toothed or entire. Flowers in leafy or bracteate terminal spikes or only in the lower axils, ± sessile, polygamous or monoecious, rarely dioecious, the upper flowers commonly ♂, the lower ♀; bracteoles 2, often very inconspicuous. Calyx inconspicuous, of 4 small lobes in the ♂ flowers, minute in the ♀ flowers. Corolla of 2–4 boat-shaped deciduous petals in the ♂ flowers, minute or absent in the ♀ flowers; stamens usually 8 or sometimes 1, 2, 4 or 6, absent in ♀ flowers. Ovary 2–4-locular, 4-ovuled, ± 4-sulcate, reduced or absent in ♂ flowers; style very short or absent; stigmas 2 or 4, subsessile, oblong, recurved, persistent. Fruit separating into 1-seeded nutlets, usually 4 or sometimes fewer by abortion; pericarp often ± tuberculate. Seeds pendulous, oblong-cylindric, with a membranous testa and copious endosperm.

A cosmopolitan genus of about 40 species. Two introduced species occur in the Flora area. *M. aquaticum* has been collected several times and since it may well spread is treated fully. A single specimen of *M. spicatum* L. has been washed up on the shore of Lake Tanganyika (Ufipa District: Kasanga, Songambele Beach, 29 Nov. 1969, *Wingfield* 528), but, since there is a specimen from Lake Nyasa collected by Laws in 1878, it may well be a species introduced by migrating birds long ago. *M. spicatum* is easily distinguishable from *M. aquaticum* by its inflorescences being distinctly terminal, with the flowers in the axils of whorls of toothed or quite divided short leafy bracts which are entirely different from the finely divided leaves subtending the flowers in *M. aquaticum*.

M. aquaticum (*Vellozo*) *Verdc.* in K.B. 28: 36 (1973). Type: Brazil, junction of R. Taguahy and R. Amazon, *Vellozo* (ubi?)

Aquatic herb, either growing in water, with most of the upper leafy parts above the water-level, or sometimes on mud; stems 0·45–1·2 m. long, simple or sparsely branched, puberulous or glabrous, with up to ¼ emerging, mostly several from a creeping rhizome. Leaves all pinnatipartite, in whorls of 4–6, 0·6–2·5(–3) cm. long, often puberulous; lobes alternate or subopposite, 15–17(–35), linear-subulate, 1·8–5(–7) mm. long, 0·13–0·3 mm. wide, ± acute,

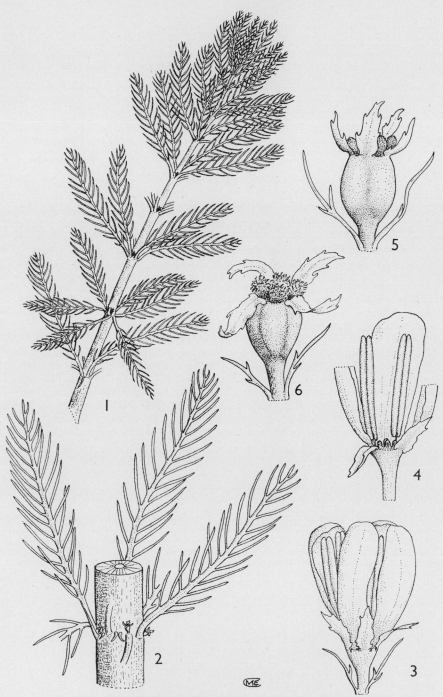

FIG. 3. *MYRIOPHYLLUM AQUATICUM*—**1,** part of leafy stem with female flowers, × 1; **2,** detail of same,
× 2; **3,** male flower, × 12; **4,** same with petal removed to show rudimentary styles, × 12; **5,** female
flower, × 24; **6,** same at a later stage, × 24. 1, 5, 6, from *Kimani* in *E.A.H.* 14302; 2, from *Hanid* 1087;
3, 4, from *Cuming* 164 (Chile). Drawn by Mrs M. E. Church.

entire or denticulate. Flowers unisexual, monoecious or mostly dioecious, solitary in the leaf-axils, the lower flowers ♀ and the upper ♂. Male flowers : bracteoles 2, linear, 1·5–3 mm. long, 0·1 mm. wide, white, very finely bifid or trifid ; pedicel and calyx-tube together at first 1·3–2·5 mm. long, tenuous, later 3·5–5 mm. long, 2 mm. thick ; lobes 4, oblong-lanceolate or narrowly triangular, 0·7–1·8 mm. long, 0·4 mm. wide, at first erect, later spreading, minutely serrate ; petals 4, boat-shaped, 2·5–5 mm. long, 0·8–1·3 mm. wide, clawed ; stamens 4–8, filaments 0·15–5 mm. long ; anthers 1·9–3·5 mm. long, 0·4 mm. wide ; sometimes minute rudiments of styles present. Female flowers : bracteoles 2, linear, 0·7–1·2 mm. long, 0·15 mm. wide, acute, sub-membranous, toothed at the middle ; pedicels 0·3–0·7 mm. long, thick ; calyx-tube 0·5–1 mm. long, 0·5–0·6 mm. wide, 4-angled, 4-grooved, the 4 lobes linear, (0·3–)0·5–1 mm. long, at first erect, later reflexed, acute, entire, minutely serrulate or pinnately ± 5-lobed ; petals and stamens lacking ; stigmas 4, short, 0·45–0·7 mm. long, at first erect, later decurved, capitate or claviform, shortly papillose. Fruit ovoid, 1·8 mm. long, 1·2 mm. in diameter, papillose-punctate, eventually dividing into 4 parts. Fig. 3.

KENYA. Nairobi R., 5 Mar. 1970, *Kimani* in *E.A.H.* 14302 ! & Nairobi Golf Course, 17 July 1967, *R. Bond* in *E.A.H.* 13810 & Nairobi National Museum garden, 24 Feb. 1970, *Mathenge* 570

TANGANYIKA. Lushoto District : E. Usambara Mts., Amani Lake, 16 Apr. 1956, *Tanner* 2750 ! & Amani, artificial pond [perhaps the lake], 26 Sept. 1967, *Harris* 1033 & same place, 2 Jan. 1968, *Dar es Salaam Univ. Students* in *DSM* 142

DISTR. **K**4 ; **T**3 ; Brazil, Peru, Uruguay, Argentina and Chile* ; naturalized or casual in parts of Europe and S. Africa

HAB. In artificial ponds and lakes, also by river banks, presumably originally introduced as an aquarium plant ; 900–1650 m.

SYN. *Enydria aquatica* Vellozo, Fl. Flumin. : 57 (1825) & Ic. Fl. Flumin. 1, t. 150 (1835
    Myriophyllum brasiliense Cambess. in A. St. Hil., Fl. Bras. Merid. 2 : 252 (1829) ;
      Kanitz in Martius, Fl. Brasil. 13 (2): 380 (1882); Schindler in E.P. IV. 225 :
      88, figs. 25, 28/k (1905) ; Cook in Fl. Europaea 2 : 312 (1968). Type : Brazil,
      not far from Sao Paulo, marshes near Jondiahi, *St. Hilaire* (MPU, holo. !)
    M. proserpinacoides Hook. & Arn. in Bot. Misc. 3 : 313 (1833) ; Kanitz in Martius,
      Fl. Brasil. 13 (2): 380 (1882). Types : Brazil, Buenos Aires, *Gillies* (GL, syn.,
      K, isosyn. !) & *Tweedie* (K, syn. !) & Chilean Andes (first and second ranges),
      Cuming 164 (GL, syn., K, isosyn. !)

   * *Ekman* 13941 from Hispaniola, determined by Urban as *M. verticillatum*, seems to be *M. aquaticum*.

INDEX TO HALORAGACEAE